The Story of the Western Australian State Fossil Emblem

by John Long

with illustrations by Jill Ruse and John Long

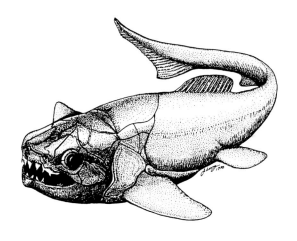

Western Australian Museum

© Western Australian Museum 2005
Reprinted April 2005
Reprinted July 2005

The National Library of Australia
Cataloguing-in-Publication entry:
Long, John A. 1957- .
Gogo fish! : the story of the Western Australian state fossil emblem.

For primary to secondary school students.

ISBN 1 920843 10 8.

1. Fishes, Fossil - Western Australia - Emanuel Range - Juvenile literature.
2. Paleontology - Western Australia - Emanuel Range - Juvenile literature.
3. Gogo Fish fossil sites - Western Australia - Emanuel Range - Juvenile literature.
I. Ruse, Jill. II. Western Australian Museum. III. Title.

567.2

Edited by Jane Hammond-Foster
Designed by Jill Ruse
Illustrations by Jill Ruse and John Long
Published by the Western Australian Museum
Francis Street, Perth, Western Australia
Printed by Percival Print, Wembley, Western Australia

Contents

4
Introduction

6
An Ancient Barrier Reef

10
How the Fossils Were Formed

12
The Discovery of the Gogo Fish

15
Preparing the Fossils

17
Palaeontological Detective Work

18
Naming the New Fossil Fish

20
Why We Study Fossil Fishes

22
About *Mcnamaraspis*

23
The Diversity of Gogo Fishes

27
The State Fossil Emblem Campaign

30
Proclamation of the State Fossil Emblem

32
Gogo Fishes Rock On

35
Glossary and Pronunciation Guide

38
Acknowledgements

Introduction

I AM A PALAEONTOLOGIST who works at the Western Australian Museum.

Palaeontology is the study of fossils, the ancient life of our planet. This includes dinosaurs, fossil fishes, shells and many kinds of invertebrates, plants, even fossilised footprints.

My particular specialty is the study of early fishes, and this is what led me to Western Australia for the first time in 1986 - to work at the world-famous Gogo fish site, so named after Gogo Station in the Kimberley.

The Gogo fish fossils were first found in the 1940s by German palaeontologist Dr Curt Teichert, who was then working at the University of Western Australia.

In the 1960s a scientist at the Natural History Museum in London, Mr Harry Toombs, developed a new technique, which enabled the bones of the fossil fishes to be etched out of limestone rock using weak acetic acid. This new method showed that unlike other fossil fishes of the same age, which were flattened in the rocks, the Gogo fishes were three-dimensionally perfect.

Dr John Long at the Gogo fish site in the Kimberley.

Onychodus

This perfect, three-dimensional skull of a Gogo fish, *Onychodus*, is 375 million years old.

The Gogo site in the Kimberley Region of Western Australia.

In 1963 and 1967 there were two large expeditions to Gogo by the British Museum, Western Australian Museum and Hunterian Museum of Scotland. They found a large number of new fossil species of fish and shrimp-like crustaceans.

In 1986 I came to Western Australia to study the Gogo fishes. I hoped to find new kinds of fossil fishes, and through my study of them gain new information on the early evolution of fishes.

This is the story of how I found one of those fossils, a Gogo fish, which I eventually named *Mcnamaraspis kaprios*. Today it is the most famous fossil in Western Australia, as it's our State fossil emblem.

John Long 2003

An Ancient Barrier Reef

375 MILLION YEARS AGO, during time known as the Devonian Period, the northern part of Western Australia was home to a huge barrier reef that teemed with a great diversity of life. The seas were warm and tropical because northern Australia was situated close to the equator.

Lloyd Hill as seen from the air was once an ancient atoll in a shallow warm sea. The Gogo fish fossils are found in the valleys which today represent the ancient sea floor.

Stromatolite

A different kind of reef

This reef stretched from near Derby some 1500 kilometres away right around to Kununurra. It fringed the Kimberley, which was probably an island. It was home to many varieties of strange fishes, squid-like animals called ammonoids and nautiloids, beetle-like trilobites, sea lilies, corals, lampshells, primitive prawns, snails and shellfishes. This reef was not built of coral like today's Great Barrier or Ningaloo Reefs, but composed of deposits formed primarily by bacteria and algae (called 'stromatolites') and extinct layered sponge-like animals (called 'stromatoporoids').

Corals

Fossil tabulate corals.

Brachiopods

Brachiopods or 'lamp shells'.

Gasteropods

A gasteropod or marine snail.

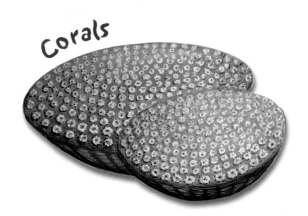

Corals

More kinds of ancient coral.

A crinoid or 'sea lily' (left), a rugose coral (right).

Trilobites

Trilobites, an extinct group of joint-legged animals.

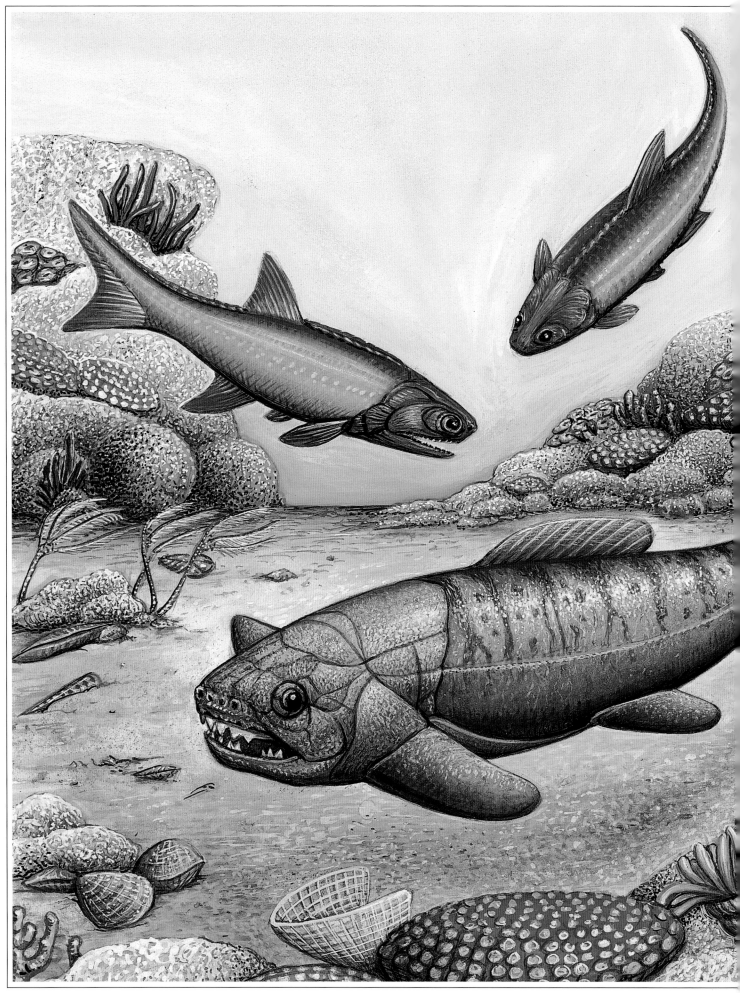

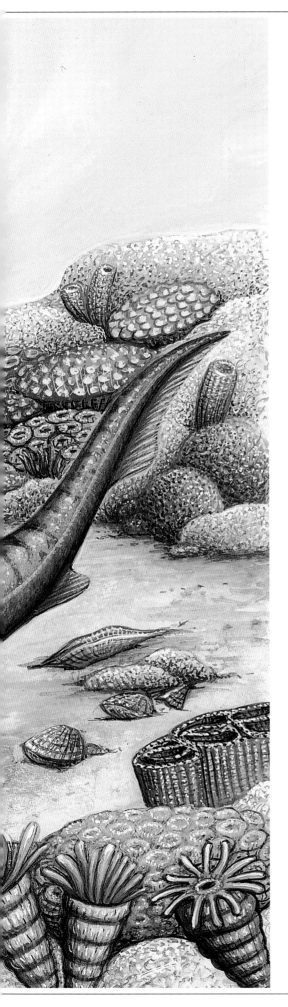

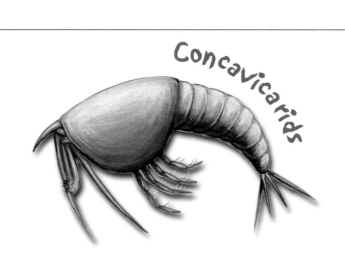

A concavicarid, an ancient shrimp-like crustacean.

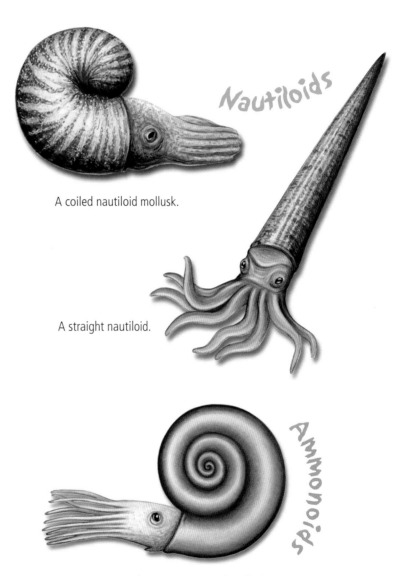

A coiled nautiloid mollusk.

A straight nautiloid.

An ammonoid, an extinct type of mollusk.

A reef scene from 375 million years ago shows the placoderm *Mcnamaraspis* (centre) hunting near the bottom of the sea whilst two ray-finned fishes, *Mimia*, swim above.

How the fossils were formed

When the fishes and shrimps living around the reef died some became buried in soft mud in the deep basins between the atolls and reef fronts. The mud around these dead animals was rich in calcium carbonate, a chemical which hardens to form limestone around the animal's remains. As the sea levels eventually subsided, the ancient reef was later buried under many layers of young sedimentary rock.

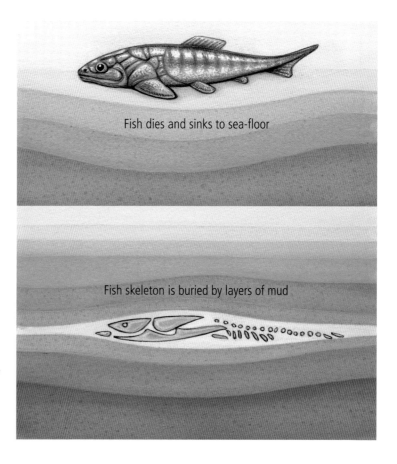

Fish dies and sinks to sea-floor

Fish skeleton is buried by layers of mud

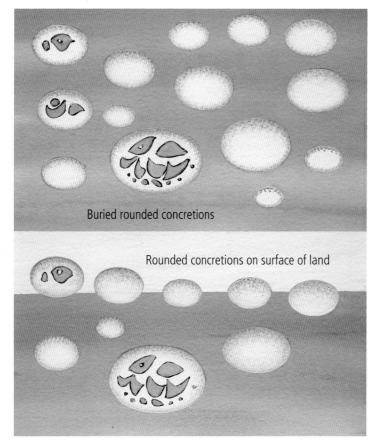

Buried rounded concretions

Rounded concretions on surface of land

After hundreds of millions of years, the effect of rain and weather, together with gradual uplift of the land, has slowly removed these surface layers and once again exposed the fossilised reef.

Rounded concretions, which formed around the marine creatures of the reef, now litter some of the valleys. These include the world-famous Gogo fish sites.

Today, the reef can be seen as a series of prominent limestone hills in the Fitzroy Crossing area. Popular tourist sites like Windjana Gorge, Geikie Gorge and Tunnel Creek are all part of this ancient reef.

Geikie Gorge on the Fitzroy River exposes 375 million year old rocks which once formed the ancient reef.

Did You Know?

Did you know that brachiopods, also known as lamp shells, take their name from the ancient Greek meaning "arm-foot"? This is because many have a pair of feathery arm-like structures, called brachidia, that were thought to be able to push the animal along.

foot

The Discovery of the Gogo Fish

IN JULY 1986, a team of volunteer workers visited the Gogo fish sites in order to search for new specimens. They drove all the way from Perth in a four-wheel drive car, a journey of some 3,800 kilometres.

Before leaving the team had written to the landowners and station managers at Gogo and Mt. Pierre Stations to get their permission to work on the land. Eventually, after four days of driving, they arrived at the fossil sites and set up camp in the limestone ranges.

During the first week at the fossil sites they spent all day breaking rocks but didn't find a single fossil fish. Then, gradually, as they searched further afield they started to find them.

For each full day's work only two or three good fish fossils were found. Only one in every thousand of the rounded rocks in the Gogo site has a fish fossil in them! Each day the team would smear themselves with sun cream, pack a bottle of water and some snacks, then go wandering out into the valleys to search for fossils.

A Western Australian Museum team searching for fossil fishes in the ancient reef deposits near Fitzroy Crossing.

They hit the rounded rocks with their hammers. If there was a fossil inside it would always split along the right layer to expose the fossil. The rocks containing fossils were carefully wrapped and taped with masking tape. The fossils would be clearly labelled with a marker pen, stating the date, the locality, who found it, and what kind of fossil it was. They were then packed into boxes with lots of wrapping around them so they would not break during the long journey back to Perth.

Gogo nodules slowly erode out of the exposed bedrock.

Then on August 1st 1986 in Bugle Gap, about 100 kilometres east of Fitzroy Crossing, Dr John Long's hammer struck a rounded rock and revealed a truly amazing fossil. It was an almost complete skeleton of a 375 million-year-old fish.

At first the fossil could only be recognized as belonging to one of the ancient armoured fishes called placoderms. Placoderms have bony plates on the outside of their bodies, which often show little rounded warts on the surface. Later that day a skull of a primitive bony fish was found. The fish had shiny bones covered by a layer of tooth-like tissue called dentine.

This was the first time a complete head of this kind of rare fish had ever been found at Gogo.

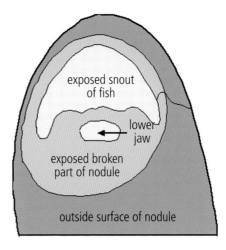

exposed snout of fish

lower jaw

exposed broken part of nodule

outside surface of nodule

Fossil lungfish head exposed after breaking a nodule. The complete skull inside this rock revealed after preparation is shown on page 24.

During the time working at the site the team encountered lots of native wildlife, including kangaroos and wallabies, many kinds of birds, snakes, lizards and bats. In Geikie Gorge storks danced on their long spindly legs; scorpions and large hairy spiders prowled around for food and at night bats would fly over the campfire.

The team spent five weeks working in the region and found many good fossil fish specimens. They would not know what kinds of fossil fishes had been found until they arrived back in Perth and started the long, slow process of carefully preparing the fossils from out of the rocks.

Above left: Black-necked Stork *Ephippiorhynchus asiaticus* (Jabiru) illustrated by J.C. Darnell/G.Miller.
Top right: A baby death adder found in the field. This one is being handled by a snake expert.
Above: A freshwater crocodile lying beside a pool at Windjana Gorge.

Preparing the Fossils

BACK IN A LABORATORY, at the University of Western Australia, the broken pieces of the rocks containing the fishes were put back together using strong glue, then immersed in a solution of 10% acetic acid. Acetic acid is the chemical found in vinegar. It will dissolve limestone rock, which is made of calcium carbonate, but doesn't dissolve bone, which is made of a different mineral called 'hydroxyapatite'.

A slow process

Slowly, day-by-day as more of the rock dissolved, the bones unaffected by the acid protruded out of the rock. Then the fragile bones were washed in fresh water to clean the acid residue out of their pores.

Once the bones had dried dilute glue was applied to them with a paintbrush. This soaked into them and hardened them from the inside.

As the Gogo fish is etched out of the limestone rock the bones are very fragile and must be hardened with dilute glue.

After two months of slowly dissolving the rock, and hardening the bones, they could be separated from the rock. Then they are carefully reassembled to form skeleton.

The particular specimen, which was found at Bugle Gap on August 1st, looked very different from all the other prepared placoderms. As placoderm plates have flat, overlapping margins, each plate can be tightly fitted next to its neighbouring plate. It's like doing a three-dimensional jigsaw puzzle. Each bone is stuck together to restore the three-dimensional form of the fish.

A Gogo lungfish jaw slowly emerges out of the rock after treatment in weak acetic acid, x 0.8.

headshield　　　　　　　　　　　　　　　trunkshield

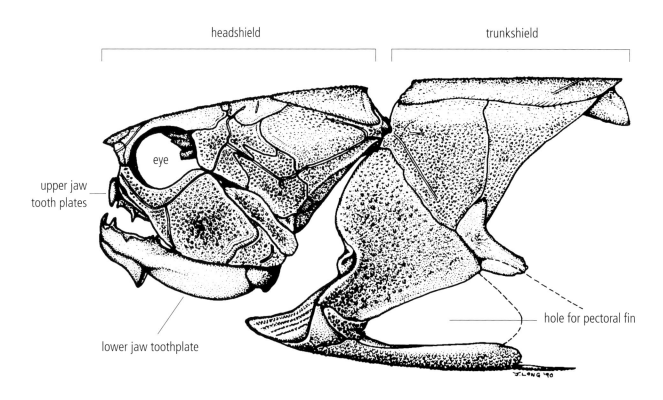

upper jaw
tooth plates

eye

lower jaw toothplate

hole for pectoral fin

Top: The fully prepared specimen of *Mcnamaraspis kaprios* showing the skeleton in side view, x 1.25.
Above: A sketch of the *Mcnamaraspis* specimen showing some of its special features.

Palaeontological Detective Work

THE NEXT JOB was to make a careful study of the new specimen. It was photographed and measured and the structure of its jaws and teeth studied. Comparisons were made with all other known placoderms in its family and the unique features showed that this was in fact a new species.

The Natural History Museum, London.

To confirm that the Gogo fish was new to science, museums were visited and letters written around the world until in 1992 there was a major breakthrough. Another example was found in the Natural History Museum in London. It was a smaller individual collected from Gogo in 1967 and had been wrongly labelled but it showed the same unique features as the Gogo fish. Now the new species could be given a species and genus name.

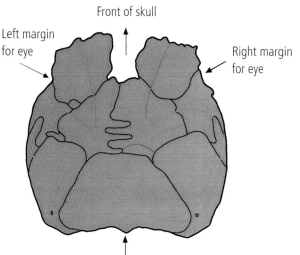

Front of skull

Left margin for eye

Right margin for eye

Nuchal or neck plate

Above: The London specimen of *Mcnamaraspis* is kept in the Natural History Museum.

Left: A placoderm fossil from Germany which is similar to *Mcnamaraspis*. Inset: A Gogo placoderm called *Compagopiscis*, which is a close relative of *Mcnamaraspis*, x 1.

Naming the New Fossil Fish

The Tiger, *Panthera tigris*, is another species of big cat similar to the Lion, *Panthera leo*.

ALL ANIMALS AND PLANTS are classified by giving them Latin or Greek names that are composed of a genus and a species name.

This means that irrespective of what common names in different languages they are given, there is a universal naming system in place. For example, the lion and tiger are closely related forms of big cat which are both in the same genus, *Panthera*.

The African lion is a different species from the tiger and is called *Panthera leo*, whereas the tiger is called *Panthera tigris*.

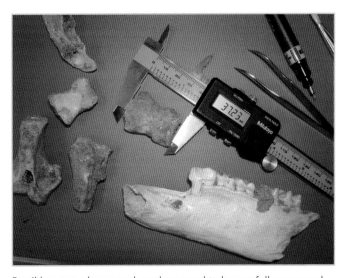

Fossil kangaroo bones such as these need to be carefully measured and studied before they can be accurately identified.

Species are defined in the simplest way as being animals and plants of similar form which can breed successfully with their own, but not with other species. When we refer to fossil species we obviously cannot test whether they were capable of breeding, so instead we base our study on the shapes and variability of their bones. By using the measurements of bones of living animal species we can identify the variations in size and shape in order to study the bones of an extinct species.

The investigations had shown that the new Gogo fish was not only a new species, but that it did not fit into any of the existing genera of placoderms, so it required a new genus name as well. The discoverer of a new species usually gets to name it. In order to do this a description of the new fish, outlining its unique characteristics, is published in an internationally recognised science journal.

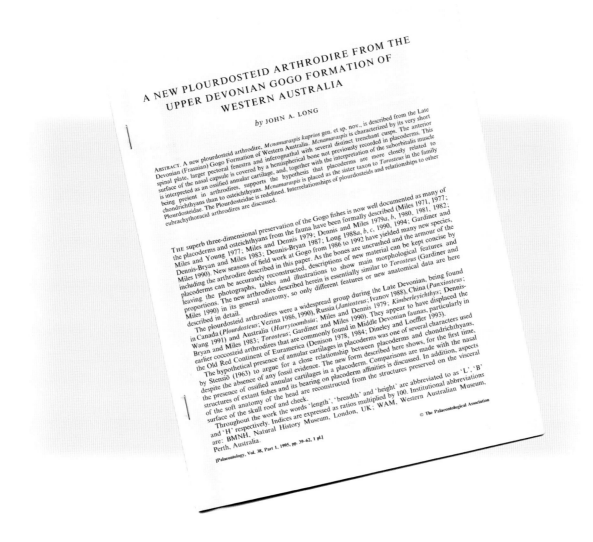

A NEW PLOURDOSTEID ARTHRODIRE FROM THE UPPER DEVONIAN GOGO FORMATION OF WESTERN AUSTRALIA

by JOHN A. LONG

ABSTRACT. A new plourdosteid arthrodire, *Mcnamaraspis kaprios* gen. et sp. nov., is described from the Late Devonian (Frasnian) Gogo Formation of Western Australia. *Mcnamaraspis* is characterized by its very short spinal plate, larger pectoral fenestra and inferognathal with several distinct trenchant cusps. The anterior surface of the nasal capsule is covered by a hemispherical bone not previously recorded in placoderms. This is interpreted as an ossified annular cartilage, and, together with the interpretation of the suborbitalis muscle being present in arthrodires, supports the hypothesis that placoderms are more closely related to chondrichthyans than to osteichthyans. *Mcnamaraspis* is placed as the sister taxon to *Torosteus* in the family Plourdosteidae. The Plourdosteidae is redefined. Interrelationships of plourdosteids and relationships to other eubrachythoracid arthrodires are discussed.

THE superb three-dimensional preservation of the Gogo fishes is now well documented as many of the placoderms and osteichthyans from the fauna have been formally described (Miles 1971, 1977; Miles and Young 1977; Miles and Dennis 1979; Dennis and Miles 1979*a*, *b*, 1980, 1981, 1982; Dennis-Bryan and Miles 1983; Dennis-Bryan 1987; Long 1988*a*, *b*, *c*, 1990, 1994; Gardiner and Miles 1990). New seasons of field work at Gogo from 1986 to 1992 have yielded many new species, including the arthrodire described in this paper. As the bones are uncrushed and the armour of the placoderms can be accurately reconstructed, descriptions of new material can be kept concise by leaving the photographs, tables and illustrations to show main morphological features and proportions. The new arthrodire described herein is essentially similar to *Torosteus* (Gardiner and Miles 1990) in its general anatomy, so only different features or new anatomical data are here described in detail.

The plourdosteid arthrodires were a widespread group during the Late Devonian, being found in Canada (*Plourdosteus*; Vezina 1986, 1990), Russia (*Janiosteus*; Ivanov 1988), China (*Panxiosteus*; Wang 1991) and Australia (*Harrytoombsia*; Miles and Dennis 1979; *Kimberleyichthys*; Dennis-Bryan and Miles 1983; *Torosteus*; Gardiner and Miles 1990). They appear to have displaced the earlier coccosteid arthrodires that are commonly found in Middle Devonian faunas, particularly in the Old Red Continent of Euramerica (Denison 1978, 1984; Dineley and Loeffler 1993).

The hypothetical presence of annular cartilages in placoderms was one of several characters used by Stensiö (1963) to argue for a close relationship between placoderms and chondrichthyans, despite the absence of any fossil evidence. The new form described here shows, for the first time, the presence of ossified annular cartilages in a placoderm. Comparisons are made with the nasal structures of extant fishes and its bearing on placoderm affinities is discussed. In addition, aspects of the soft anatomy of the head are reconstructed from the structures preserved on the visceral surface of the skull roof and cheek.

Throughout the work the words 'length', 'breadth' and 'height' are abbreviated to as 'L', 'B' and 'H' respectively. Indices are expressed as ratios multiplied by 100. Institutional abbreviations are: BMNH, Natural History Museum, London, UK; WAM, Western Australian Museum, Perth, Australia.

© The Palaeontological Association

[Palaeontology, Vol. 38, Part 1, 1995, pp. 39–62, 1 pl.]

The paper describing the Gogo fossil was sent to the British journal *Palaeontology* in November 1993, and officially accepted for publication in May 1995. This was the date that the fish was officially named as *Mcnamaraspis kaprios*.

The Greek ending '*aspis*' means shield, so the literal translation means 'McNamara's Shield'. The species name '*kaprios*' was Greek for boar-like, as the feisty little fish had prominent boar-like tusks on its lower jaws.

The name *Mcnamaraspis* was a tribute to the work of Dr Ken McNamara of the Western Australian Museum who has done an extraordinary amount of research on a variety of Western Australian fossils.

Dr Ken McNamara of the
Western Australian Museum.

Why We Study Fossil Fishes

MANY PEOPLE ask what is the value of studying fossil fishes, and how does it benefit our society?

The main contribution is to understand the early evolution of fishes. Evolution is the gradual change in form, from one species into another species, over prolonged time.

There are many ways we can study evolution. By looking at the fossil record we can see how ancient species of animals changed into later, more specialised species. Scientists who study the biology of living species of animals and plants can measure their degree of evolutionary change through various methods, such as comparing their DNA (deoxyribonucleic acid), a complex molecule unique to every species, in fact, unique to some degree in every living organism.

Back in the Devonian Period, often called the age of fishes, fishes were the highest evolved back-boned animals. By the end of that period the first four-legged animals, the amphibians, had evolved from some of these fishes. By studying the Gogo fish fossils we are able to see how these changes in fish skeletons compare to our own human skeleton.

Gogo fishes are particularly important because they are so well preserved and their anatomy provides detailed information, which helps us understand the various stages in fish evolution and explore the relationship between fossil fishes and all living animals.

Did you know?

That the pattern in our human arm bones first appeared within Devonian fishes? The most primitive fish known to have a true humerus (upper arm bone) is *Onychodus* from Gogo (see page 25).

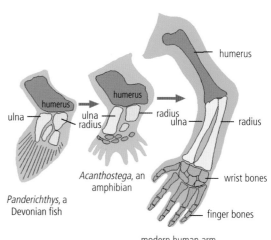

Panderichthys, a Devonian fish

Acanthostega, an amphibian

modern human arm

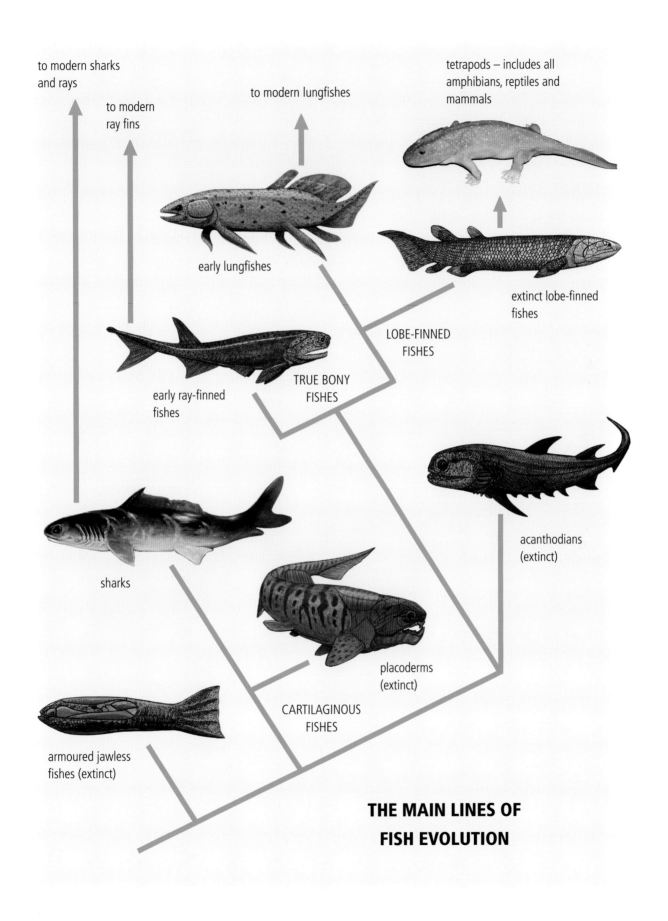

to modern sharks
and rays

to modern
ray fins

to modern lungfishes

tetrapods – includes all
amphibians, reptiles and
mammals

early lungfishes

extinct lobe-finned
fishes

LOBE-FINNED
FISHES

early ray-finned
fishes

TRUE BONY
FISHES

acanthodians
(extinct)

sharks

placoderms
(extinct)

CARTILAGINOUS
FISHES

armoured jawless
fishes (extinct)

**THE MAIN LINES OF
FISH EVOLUTION**

About Mcnamaraspis

*M*CNAMARASPIS was a voracious placoderm fish about 30cm long. It had broad front fins, and although the tail was not preserved, from the study of other similar placoderms it was almost certainly shark-like. The front of the fish's body, including the entire head was covered in a shield made of interlocking bones, protecting it from its enemies.

The main distinguishing features of *Mcnamaraspis* are its tiny spinal plates, which shoulder the large front fins, and its characteristic jaws with prominent pointed tusks. These features suggest that it used its sharp teeth for hunting other fishes. Its mouth was able to open wide, and a small-toothed bone on the inside of its palate helped it to secure its prey.

Scientists had debated for many years which other fishes were related to placoderms. Were they closely allied to the modern bony fish like salmon or trout? Or were they relatives of the cartilaginous fishes like sharks and rays?

The most exciting discovery concerning *Mcnamaraspis* was that this fish was the first placoderm ever found to have its nose area extremely well preserved. Inside its snout were delicate little bones called 'annular cartilages' that showed how similar it was to living sharks.

The study of *Mcnamaraspis* also enabled a reconstruction of some parts of the soft anatomy of its head. Because Gogo fishes bones are so well preserved the position of certain soft tissues can be determined from the scars and grooves underneath the skull. That is, how the muscles of the jaws and eyes are attached to the skull, and how some of the gill arches worked, enabling the fish to breathe in water.

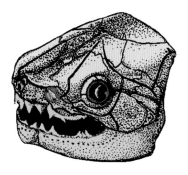

In addition to studying the anatomy of *Mcnamaraspis*, a study of closely related placoderm fishes was made, to indicate which other fishes were its closest relatives. These turned out to be other Gogo placoderms named *Harrytoomsia* and *Goujetosteus*.

The restored head of *Mcnamaraspis* showing its nasal capsules at the front of its snout. Red arrows show the incurrent and excurrent nostrils.

The Diversity of Gogo Fishes

THE SEA in which *Mcnamaraspis* lived was also home to a great variety of other fishes. Most common are placoderms. More than 25 different species of them have been identified.

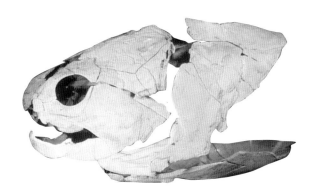

The biggest of these was *Eastmanosteus*, a three-metre long shark-like predator that may have hunted *Mcnamaraspis*.

Eastmanosteus

There were several other small predatory placoderms in the same family group as *Mcnamaraspis*. *Goujetosteus, Kimberleyichthys* and *Harrytoomsia*, the latter named after Mr Harry Toombs of the Natural History Museum in London who prepared the first Gogo fishes.

Some large placoderms, like *Holonema*, which grew to 1.5 metres long, may have sifted the sea floor for algal balls. Its teeth are not sharp like those of *Mcnamaraspis*, but spoon-shaped with many fine ridges.

Specimens of *Holonema* have been found with tiny pebbles inside the gut, possibly cores of algal balls called 'oncolites' that lived on the front slope of the ancient reef.

Holonema

The most bizarre-looking placoderm at Gogo is *Bothriolepis*, a box-like fish with long segmented bony arms. *Bothriolepis* probably dug itself into the muddy sea floor and used its weak jaws for sifting the sediment for worms and rotting algae.

Bothriolepis x 0.5

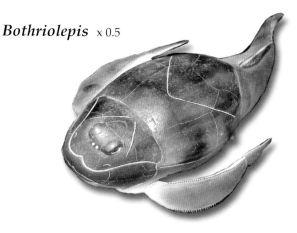

Campbellodus x 0.5

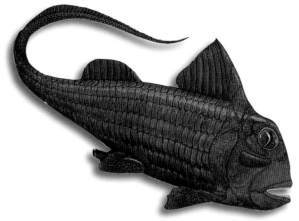

Other placoderms, like *Campbellodus,* had a short, deep head with very large eyes and powerful crushing tooth plates. It may have flitted about the sea floor, crunching up clams and snails.

There were many kinds of lungfishes living in the Gogo seas. Long-snouted forms such as *Griphognathus* looked more like ducks with fins. These nuzzled around the muddy sea floor searching for juicy worms, while others with hard tooth plates in their mouths, such as *Chirodipterus*, crunched up hard-shelled prey.

Chirodipterus

Griphognathus x 0.2

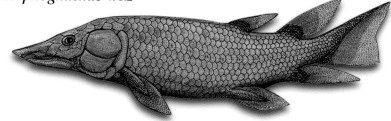

Others like *Holodipterus* had very strong jaws with the entire roof of the mouth covered with many small teeth.

Holodipterus   x 0.33

Perhaps the most ferocious of all the bony fishes at Gogo was the dagger-toothed fish called *Onychodus*. At 2 metres long, its eel-like body enabled it to dart out of crevices in the reef and catch unwary, slow fish. It had two huge rows of curved dagger-like teeth at the front of its mouth, which could rotate outwards whenever it opened its large mouth.

Onychodus

One specimen from Gogo had the remains of its last meal still stuck in its gullet. The placoderm it had caught would have been about 30cm long, and the specimen of the *Onychodus* was estimated at about 60cm long. It may have choked on a meal that was too big for its belly.

Gogonasus   x 1

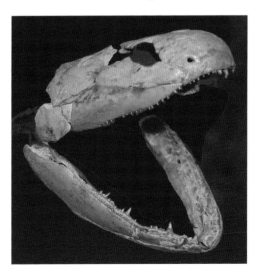

One of the most exciting discoveries at Gogo, also made on August 1st 1986, was the first complete skull of an osteolepiform fish from Gogo. *Gogonasus* (meaning 'snout from Gogo') provided lots of detailed information about the anatomy of these fishes, the group ancestral to the first land animals.

The fossil ray-finned fishes from Gogo are commonly found as complete individuals. Two forms are known, *Mimia* and *Moythomasia*. These probably swam in shoals in the open waters catching smaller fishes and tiny floating worm-like creatures called conodonts.

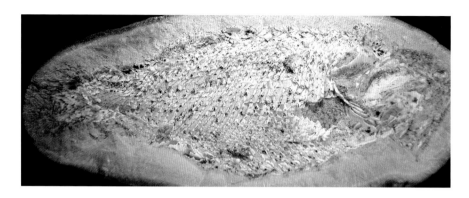

Moythomasia x 0.75

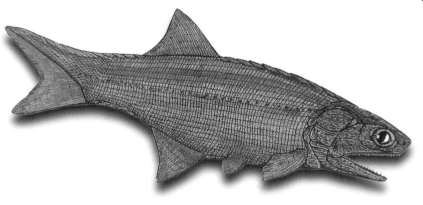

Mimia x 1

Since collecting at Gogo began in earnest in the 1960s, scientists have identified over 40 different species of prehistoric fishes, all of which were new to science. Gogo now ranks as the most diverse and well-preserved fish fauna site of its age from anywhere in the world.

The State Fossil Emblem Campaign

THE CAMPAIGN for a State Fossil Emblem began in March 1994 when a Sutherland Primary School parent, Carolyn Symmonds, was reading a book called *The Big Golden Book of Dinosaurs* to one of her children.

She read a dedication in that book to the girls and boys of McElwain Elementary School in Denver, USA, and to Ruth Sawdo, the teacher whose fourth grade students led a spirited and successful lobby to make the dinosaur *Stegosaurus* the Colorado Official State Emblem.

Mrs Symmonds suggested to the school's principal, Mr John Shorthill, that the school organise a similar campaign for a Western Australian State Fossil emblem. Western Australia already had the black swan; the numbat and the kangaroo paw as our bird, mammal and floral emblems.

The school's main aim in promoting the campaign was to take full advantage of its educational potential, not just for its own students but also for the general public.

The school's work on the campaign gave both students and teachers the chance to learn much about fossils, Western Australia's prehistory and evolution. It also gave students the opportunity to understand the political process and role of lobbying in a democratic society.

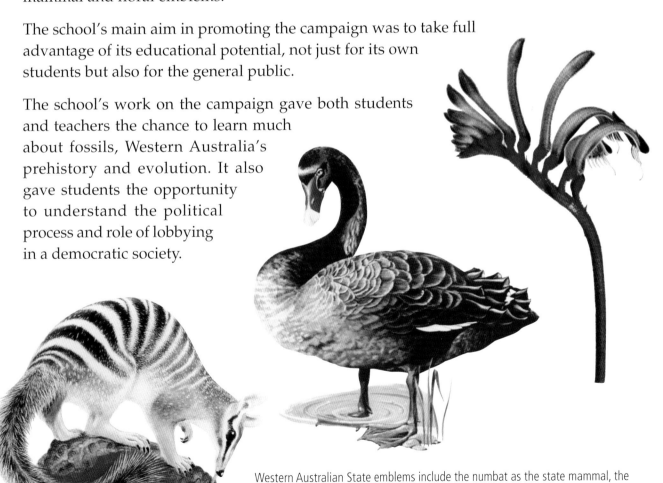

Western Australian State emblems include the numbat as the state mammal, the black swan as the state bird, and Mangles kangaroo paw as the state flower.

The School's Parent and Citizen's Association and the Western Australian Museum were enthusiastic about the idea. Mrs Debra Parry, a teacher at the school, was appointed to coordinate the campaign.

After deciding that a Gogo fish would make a suitable emblem, the pupils researched details of different species. A vote was then taken on what the pupils thought would be the best emblem and they chose *Mcnamaraspis kaprios*.

Following lobbying of the government by the pupils, the Government accepted the merit of a State fossil emblem, and the campaign was officially launched with the Premier, Richard Court, calling for submissions.

A committee of local palaeontologists and geologists was appointed to evaluate the proposals. Public submissions were invited over the course of two months and members of the public could write to the committee, giving their reasons why they thought a particular fossil would be the best for a State fossil emblem.

On October 26th 1994, the children of Sutherland Primary School delivered their 219-page submission to Education Minister Norman Moore and asked for support from Arts Minister Peter Foss. Several other submissions were received which included nominations for a Permian brachiopod shell, a Devonian ammonoid, a Silurian trace fossil, and a Cretaceous marine reptile.

Minister for the Arts, Honorable Peter Foss, shows the Gogo fish specimen to children of Sutherland Primary School in October 1994 after announcing support for having a state fossil emblem. Courtesy The West Australian. Photo by John Mokrzycki.

Submissions

Six submissions favoured fossil stromatolites and four different species were suggested. One submission nominated the living stromatolites of Shark Bay, but this was not counted, as it didn't represent a fossil.

The most support received for any single fossil came from the Sutherland Primary School campaign. Their proposal included letters from fifteen Western Australian schools, totalling 745 signatures. In addition their petition included another 112 names in letters of support from several major museums and universities from Australia as well as North America, Japan, England and Germany.

A fossilised stromatolite, 3.5 billion years old, from Western Australia.

Children's book authors Paul Jennings and Morris Gleitzman also supported the Gogo fish and Mr Bob Halligan, who had consulted with various members of the Geological Society of Australia, included a separate submission giving strong support for the Gogo fish nomination.

Did you know?

Over 25 states in North America have official state fossil emblems. Some of these include the dinosaurs *Allosaurus* (Utah), *Stegosaurus* (Colorado), *Maiasaura* (Montana) and *Triceratops* (Wyoming and South Dakota), a dinosaur footprint (Connecticut), a fossil whale (Alabama and Mississippi), a sea scorpion (New York State), and fossil wood (North Dakota, Arizona and Louisiana).

Triceratops, the official State fossil emblem of Wyoming and South Dakota, USA.

Proclamation of the State Fossil Emblem

ON DECEMBER 5TH 1995, the Governor of Western Australia, His Excellency Major General Philip Michael Jeffrey, proclaimed the Gogo fish *Mcnamaraspis kaprios* as the official state fossil emblem of Western Australia.

A ceremony was held at Sutherland Primary School attended by the Premier, the Minister for the Arts, the Minister for Education, and the scientists involved in its discovery.

On September 4th 1997, Australia Post issued a stamp with a reconstruction of *Mcnamaraspis* on it, painted by Sydney artist Peter Schouten. It was the first time in Australian history that an extinct species of fish had adorned a postage stamp.

The most complete specimen of *Mcnamaraspis*, found on August 1st 1986, is on display in the Western Australian Museum's *Diamonds to Dinosaurs* Gallery.

State fossil emblem of Western Australia
Mcnamaraspis kaprios
This placoderm fish, named after palaeontologist Ken McNamara, was an active predator on the ancient Devonian reef of the Kimberely. It was found 100 km east of Fitzroy Crossing in 1986, and prepared out of its limestone encasement by weak acetic acid. The remains shown here comprise most of the head and trunk skeleton, with the tail missing. In December 1995 it was proclaimed by the Governor General as the official State fossil emblem of Western Australia as a symbolic representation of the rich geological heritage of the state.

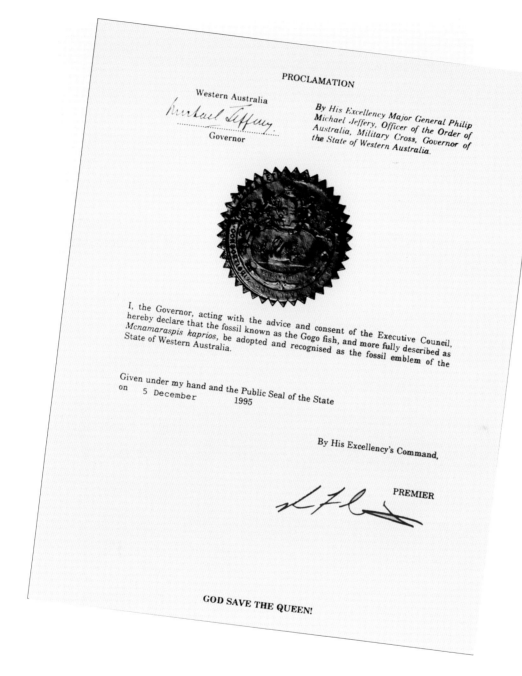

PROCLAMATION

Western Australia

Michael Jeffery

Governor

By His Excellency Major General Philip Michael Jeffery, Officer of the Order of Australia, Military Cross, Governor of the State of Western Australia.

I, the Governor, acting with the advice and consent of the Executive Council, hereby declare that the fossil known as the Gogo fish, and more fully described as *Mcnamaraspis kaprios*, be adopted and recognised as the fossil emblem of the State of Western Australia.

Given under my hand and the Public Seal of the State on   5 December   1995

By His Excellency's Command,

PREMIER

GOD SAVE THE QUEEN!

The official proclamation of the Western Australian State fossil emblem.

Left: The specimen of *Mcnamaraspis* on permanent display at the Western Australian Museum.

Gogo Fishes Rock On

THE GOGO FISH SITES are considered one of the world's most important fossil sites. British nature presenter Sir David Attenborough chose the Gogo site to talk about fish evolution for his series *Life on Earth*.

The Gooniyandi Aboriginals of the Mimbi Community who live on the land where the Gogo fossils are found are protecting the sites and promoting tourism to the region. This is so that everyone can visit and learn about the fossils and about Aboriginal heritage. They prevent unauthorised people from collecting important fossils, which could be lost to science. It is important that everyone can enjoy the Gogo fish fossils by looking at them in museums.

An aboriginal painting of an echidna from a cave overlooking the Gogo fish deposits. With permission from local land owners.

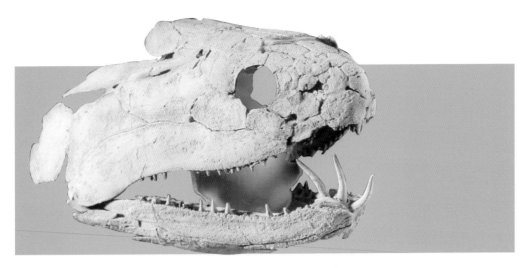

The skull of *Onychodus* from Gogo, on display at the Western Australian Museum, x 0.75.

There are now Gogo fish on display in some of the major natural history museums in England, America, Canada and Russia, through official exchanges with the Western Australian Museum.

But this isn't the end of the Gogo fish story. More Gogo fishes are being discovered every year as teams of scientists continue their work in the area. The 2001 expedition to the area discovered another four new species of Gogo fish, including lungfishes, placoderms and a new type of ray-finned fish. My children have joined me on my field trips and helped find new fishes. One is named after my eldest daughter, Sarah, *Gogosteus sarahae* (meaning Sarah's Gogo bones).

I hope that one day some of you will become palaeontologists, and go out and find your own new species of fossil. I'm sure that there are plenty left out there to discover.

The skull of *Gogosteus sarahae* named after Sarah Long, x 0.75.

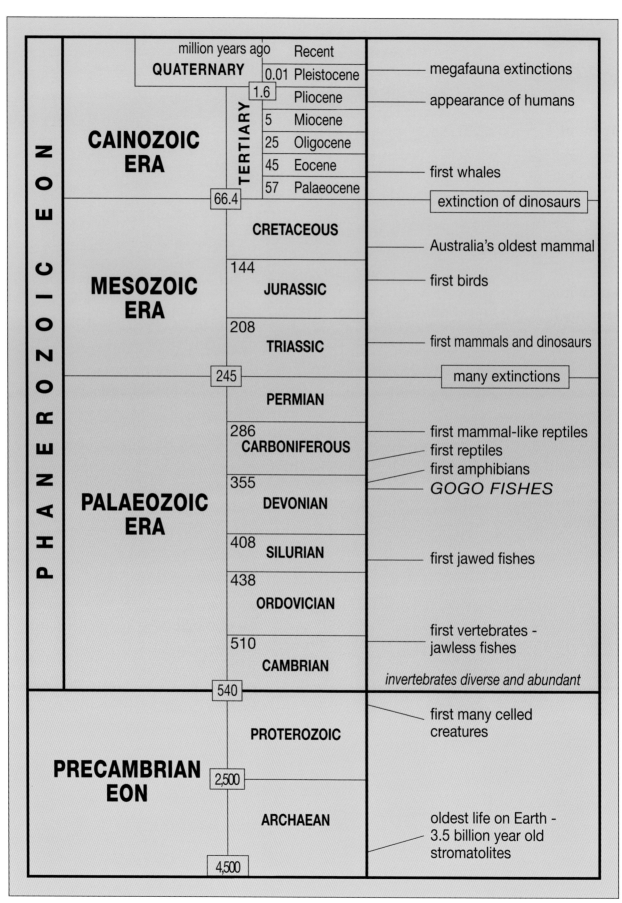

The geological time scale.

Glossary and Pronunciation Guide

actinopterygian (ack-tin-op-ter-ridge-ee-an): ray-finned bony fishes, e.g. trout, tuna or goldfish.

ammonoid (am-mon-oyd): a squid-like creature that lived inside a hollow coiled shell.

annular cartilage (an-you-lar cart-il-aj) small cartilage or bone in the front of the snout in placoderms and sharks, probably to direct flow of water around the nasal capsules.

Bothriolepis (both-ree-o-leap-iss): a box-like placoderm with segmented bony arms.

brachiopod (brak-ee-o-pod): a lamp shell, having two unequal valves, one with a small hole in it for attaching the shell to its substrate.

Campbellodus (cam-bell-o-duss): a short-shielded placoderm with a high spine on its back and large crushing tooth plates.

Chirodipterus (kire-o-dip-ter-us): a short-snouted lungfish with crushing tooth plates.

chondrichthyan (con-drick-thee-an): cartilaginous fishes such as sharks, rays and chimaerids. They lack bone in their skeletons.

crustacean (crust-ay-see-an): crabs, prawns, and their kin.

Devonian Period: from 355 million years ago until 410 million years ago, often referred to as the 'age of fishes'.

Eastmanosteus (east-man-os-tee-us): a large predatory placoderm that grew to 3 metres long.

Griphognathus (gry-fog-nath-uss): a long-snouted (almost duck-billed) lungfish that grew to 1 metre long.

Harrytoombsia (Harry-toom-see-a): a placoderm similar to *Mcnamaraspis*.

Holodipterus (hol-o-dip-ter-us): a deep-headed lungfish with a palate covered in dentine or tubercles.

Holonema (hol-o-nee-ma): a large placoderm with scoop-shaped tooth plates, it most likely ate algal balls called oncolites.

Kimberleyichthys (kim-ber-lee-ick-thees): a placoderm similar to *Mcnamaraspis*.

Mcnamaraspis kaprios (*mac-nam-ar-as-pis + cap-ree-os):* the State Fossil Emblem, a Gogo placoderm fish.

Mimia (mee-mee-a): a small ray-finned fish from Gogo. Its name is after the 'mimi', Aboriginal spirits that lived in the rocks.

Moythomasia (moy-tom-as-ee-a): a small ray-finned fish from Gogo.

lungfish: one of the fleshy-finned fish groups (also called dipnoans), so named because they have an internal lung and can gulp air to supplement breathing through their gills.

oncolites (on-co-lites): rounded balls of algae that usually form around a grain of sand or tiny pebble.

Onychodus (on-ick-o-duss): the dagger-toothed fish from Gogo, which grew to about 2 metres long.

onychodontiform (on-ick-o-don-tee-form): an extinct group of fleshy-finned fishes with large rotational tooth whorls at the front of each lower jaw.

osteolepiform (os-tee-o-leap-ee-form): an extinct group of fleshy-finned fishes with powerful front fins and other anatomical features that indicate they were an ancestral group that gave rise to the first amphibians.

placoderm (plack-o-derm): primitive jawed fishes having interlocking bony plates covering the head and trunk regions.

reef: a large marine structure built up by the remains of living organisms.

sarcopterygian (sark-op-ter-ridge-ee-an): fleshy-finned bony fishes, which include two major groups – the lungfishes – and the coelacanths and their allies.

stromatolite (strom-at-o-lite): microbial community of cyanobacteria and algae (as in sea-weed, but here as microscopic cells of algae) which form structures by cementing floating particles in the water together.

stromatoporoid (strom-at-o-por-oyd): an organism similar to a sponge or coral that builds up a layered structure made of calcium carbonate.

trilobite (try-low-bite): an extinct group of three-segmented joint-legged marine animals (related to modern insects, crabs and spiders).

Acknowledgements

The state fossil emblem was found at Gogo through support of the National Geographic Society of America (grant #3364-86). For long-term support of my work at Gogo, I thank all my field crews and colleagues over the years, particularly Lindsay Hatcher, Mark Norton, Sarah, Peter and Maddy Long.

I thank my colleague Dr Ken McNamara for his comments on this manuscript. I would also like to thank the various teachers and children of Sutherland Primary School, Dianella, for their great idea and for seeing it through to the end.

Thanks also to Jane Hammond-Foster for editing the book, Jill Ruse for her great artwork, Dr Sergei Pisarevsky of the University of Western Australia for the Devonian world map, and to Greg Jackson and Ann Ousey, Publications Department Western Australian Museum.

WESTERN AUSTRALIA
Be touched by nature

Notes